KNOW YOUR TRAINS

James Race

Old Pond Publishing

First published 2012

Copyright © James Race, 2012

James Race has asserted his right under the Copyright,
Designs and Patents Act, 1988, to be identified as the Author of this Work

ISBN 978-1-908397-13-3

Published by:
Old Pond Publishing Ltd
Dencora Business Centre
36 White House Road
Ipswich IP1 5LT
United Kingdom

www.oldpond.com

Book design and layout by Liz Whatling
Printed and bound in China

CONTENTS

1. Alexander Pacer
2. Alstom Adelante
3. Alstom Coradia 1000
4. Alstom Juniper
5. Alstom Pendolino
6. Alstom Prima
7. Bombardier Electrostar
8. Bombardier Turbostar
9. Bombardier Voyager
10. BREL 1987 Electric Locomotive
11. BREL EMU
12. BREL Express Sprinter
13. BREL Inner Suburban EMU
14. BREL InterCity 125
15. BREL Networker
16. BREL Outer Suburban EMU
17. BREL Sprinter

18. BREL Wessex Express
19. British Rail 1961 Diesel Locomotive
20. Brush 1957 Diesel Locomotive
21. Brush 1962 Diesel Locomotive
22. Brush 1989 Diesel Locomotive
23. Brush 1993 Electric Locomotive
24. Brush 1997 Rebuilt Locomotive
25. English Electric 1952 Diesel Shunter
26. English Electric 1957 Diesel Locomotive
27. English Electric 1958 Diesel Locomotive
28. English Electric 1960 Diesel Locomotive
29. English Electric 1962 Electro-Diesel Locomotive

30. English Electric Deltic
31. GEC-Alsthom TGV-TMST Eurostar
32. General Electric InterCity 225
33. General Electric PowerHaul
34. General Motors JT42CWR
35. Hitachi Javelin
36. Hunslet Suburban EMU
37. Leyland Pacer
38. Leyland Super Sprinter
39. Met-Cam Ex-Tube Train
40. Met-Cam Super Sprinter
41. Parry People Mover PPM60
42. Peppercorn A1
43. Pressed Steel Suburban
44. Siemens Desiro

Acknowledgements

I would like to thank Syd Eade,
and everyone else who has helped in producing this book.

Picture Credits

Syd Eade 5, 10, 11, 13, 14, 15, 18, 19, 22, 23, 25, 27, 29, 30, 32, 34, 37, 38, 39, 40, 42, and 43. Richard Alger 20 and 33.

All other photographs are from the author's own collection.

Author's Notes

Names and models of train types are included for identification purposes only.
All information is given in good faith and should be used as a guide.
The author cannot be held responsible for any errors.

FOREWORD

This book shows the wide range of different trains to be found on the railways of Great Britain today. All the trains featured can be found in active service on the National network.

While some types travel widely throughout the country, others stay within certain defined areas. Obviously electric trains cannot stray where there is no current, but even the versatile diesels can sometimes be limited to working in specific areas.

I have tried to keep the book as up-to-date as possible at the time of publication, but it must be mentioned that in today's privatised world, liveries and operators can come and go very quickly. All trains are currently given 'Class' numbers to help distinguish the types, plus a three figure fleet number which can be found displayed on the train. With the multiple unit passenger trains this fleet number refers to the current formation and can change should the make-up of the set alter; however, the side numbers on the individual vehicles remain constant. The 'Class' numbers used are shown on the left-hand side of each page.

JAMES RACE
Lowestoft 2012

Alexander Pacer

Type of Train:
Diesel Multiple Unit (DMU)

Class Types:
143/144

This design forms part of the 'Pacer' family of units built for local and secondary services. Walter Alexander supplied the bodywork on either Andrew Barclay & Sons or BREL underframes.

These units were originally fitted with Leyland TL11 engines and self-changing gears transmission which was later replaced by the current Cummins LTA10-R engine and Voith gearing.

Introduced in the mid 1980s, most of these units remain in service albeit refurbished. The majority are two-car units; however, 144 014–023 have centre cars owned by West Yorkshire PTE.

Alstom Adelante

Type of Train:
Diesel Multiple Unit (DMU)

Class Types:
180

Between 2000 and 2001, 15 Adelante units were constructed for First Great Western express services.

Each of the five cars forming one unit is powered by a Cummins QSK19 with Voith hydraulic transmission producing a top speed of 125 mph.

Although built for First Great Western, they were quickly displaced and some can now be found on the east coast with other operators such as First Hull Trains and Grand Central.

Alstom Coradia 1000

Type of Train:
Diesel Multiple Unit (DMU)

Class Types:
175

Description

First entering service in 2000, these units were built for First North Western but are now operated by Arriva Trains Wales.

Power is supplied by a Cummins N14 engine with Voith hydraulic transmission under each car; this gives the train a top speed of 100 mph.

Coradia 1000s are formed into two- or three-car units, with the end carriages labelled A and C, irrespective of centre coach B being there or not. There are 134 seats on two-car units while the centre carriage takes that number up to 206 on the three-car sets.

Alstom Juniper

Type of Train:
Electric Multiple Unit (EMU)

Class Types:
334/458

Description

The third rail version of the Alstom Juniper entered service from London Waterloo in 1998, followed the next year by an overhead design operating around Glasgow.

Power is supplied by Alstom's ONIX 800 traction motors, which operate on three-phase AC (Alternating Current), but the way this is achieved with DC (Direct Current) input is complicated.

Both versions were constructed at Washwood Heath, Birmingham at the works previously belonging to Metro-Cammell although the steel body shells were fabricated in Spain.

5

Alstom Pendolino

Type of Train:
Electric Train Set

Class Types:
390

Description

Virgin Trains' Class 390s are some of the fastest domestic units operating in Great Britain. They have a top speed of 140 mph, though they are currently restricted to 125 mph when in service.

Pendolino comes from the Italian 'pendolo' meaning pendulum. The name is used for a group of 'tilting train' designs made by Fiat Ferroviaria and their successor Alstom. These units have Alstom traction motors and Dellner couplers, hidden behind a retracting front panel.

The original order was for 53 units, with eight vehicles each. A ninth was added later and some of the class are expected to become eleven-car sets.

Alstom Prima

Type of Train:
Diesel Locomotive

Class Types:
67

The Alstom Prima JT 24HW-HS locomotives were constructed in Spain using General Motors components.

Thirty of the class were supplied to English Welsh and Scottish Railways (EWS) during 1999 and 2000 to replace the ageing locomotives on Royal Mail trains. However, this work ceased when these locomotives were around five years old.

Two of the type, 67 005 and 67 006, were repainted in 'Royal Claret' for use on the Royal Train. 67 029, has been painted silver and named 'Royal Diamond' and shares in these duties. They are fitted with EMD motors, electro-pneumatic brakes and electric train heating. They are capable of a 125 mph top speed.

Locomotives 67 012 through 67 015 are fitted with remote fire suppression equipment allowing them to be operated from a driving trailer. Some have been fitted with radio signalling equipment for use on ScotRail sleeper services.

Bombardier Electrostar

Type of Train:
Electric Multiple Unit (EMU)

Class Types:
357/375/376/377/378/379

Description

These units form part of Bombardier's 'Star' family of trains which are three or four cars in length. They were first introduced in 1999 and have a modular body structure to which various interior options can be fitted.

Power is received at either 25 kV AC, from overhead wires, or 750 V DC, from a third rail, and fed into two of Bombardier's own motors per power car.

Some Electrostars are only equipped with the DC third-rail system while others have only AC pantographs. A small number are fitted with both systems and are used, for example, on through services across London. The DC-only units are designed with provisions to easily retro-fit pantographs and have a cut-out in the roof where the installation would be situated.

Bombardier Turbostar

Type of Train:
Diesel Multiple Unit (DMU)

Class Types:
170/171/172

Description

Turbostars, as the name suggests, are part of Bombardier's 'Star' family of designs. Body structures are made from extruded aluminium alloy and have bolt-on steel ends.

These units have MTU diesel engines with either Voith or ZF transmissions. While these are all similar multiple units they can only be joined with sets that have couplers of the same design.

Turbostars can have two, three or four carriages with many different interior options, although the most obvious difference is the front-end design when corridor connections are fitted.

Bombardier Voyager

Type of Train:
Diesel-Electric Multiple Unit

Class Types:
220/221/222

Description

There are three similar designs in the 'Voyager' family: the original Voyager, the Super Voyager and the Meridian/Pioneer.

All variants have a Cummins QSK19 diesel engine powering two Alstom Onix electric traction motors per carriage.

An interesting difference between the Super Voyager and other members of the family is that it is fitted with a tilting system which allows it to take corners at faster speeds. All Voyagers have a top speed of 125 mph and are capable of very rapid acceleration.

Description

BREL 1987 Electric Locomotive

Type of Train:
Electric Locomotive

Class Types:
90

Class 90s were built by British Rail Engineering Limited (BREL) at Crewe Works. They were based on the Class 87 design and intended to replace old West Coast Main Line electric locomotives.

The Class 90s have 1,250 hp traction motors with air and rheostatic brakes. These electric locomotives are fitted with time-division multiplex wiring systems, which allows them to be utilised as a push-pull locomotive using a driving trailer at the other end.

When British Rail was sectorised, locomotives 90 001 through 90 015 were used by InterCity on express passenger services, while 90 016 through 90 020 were dedicated to postal service and 90 021-024 were operated by Railfreight Distribution, with the final 26 allocated to freight duties and renumbered from 90 025-050 to 90 125-150. This denoted a lower maximum speed; their electric train supply had been isolated along with other modifications.

BREL EMU

Type of Train:
Electric Multiple Unit (EMU)

Class Types:
320/321/322/456

These units were built in York and were first introduced in the late 1980s with the third rail sets following in the early 1990s.

Most of these units have three or four coaches and have overhead pantographs for AC current collection powering four Brush traction motors in one of the centre carriages. The third-rail variants have only two carriages and two GEC DC traction motors which were recovered from older scrapped units.

Most have been refurbished over the years, and many re-allocated to different work than that for which they were originally built. The most travelled are the five 322 sets, which were new for Stansted Express services. A varied couple of years followed replacement on these services before they moved to Scotland for North Berwick services. That work has also finished and they have moved to Yorkshire.

BREL Express Sprinter

Type of Train:
Diesel Multiple Unit (DMU)

Class Types:
158/159

Description

These units were built in Derby for Regional Railways' longer distance routes. They first entered service in 1989.

Each carriage has a Cummins diesel engine and Voith transmission powering one of its two bogies, giving a top speed of 90 mph.

Two- and three-car units have been built and some two-car units have since been reformed into three-car sets with the centre driving cab out of use. These were the first DMUs built with air-conditioning equipment and do not have opening windows, although each car has an 'emergency' pair of windows that can be opened with a key by the guard should any problems arise.

Description

BREL Inner Suburban EMU

Type of Train:
Electric Multiple Unit (EMU)

Class Types:
313/314/315/507/508

These three- or four-car units were built at York and entered service from the late 1970s.

Each unit is powered by four GEC or Brush traction motors with both disc and rheostatic braking.

These local units are very basic and have no first-class seating or toilets. They can collect power from either overhead AC wiring or third rail DC; the Class 313s are unique in having both systems. However, those 313s transferred to the south coast have had their pantographs removed as part of the refurbishment programme.

BREL InterCity 125

Type of Train:
High Speed Diesel Train

Class Types:
43

The InterCity 125 high speed train (HST) was introduced in the mid 1970s and has a power car at each end of its formation. The number and type of intermediate carriages is variable.

When new, the power cars were fitted with Paxman Valenta engines, originally designed for marine use; however, all have been replaced in the last few years with more modern units, which are both cleaner and quieter.

The main use for the HSTs still centres on the work they were originally built for: travelling long distances from London to the West of England and South Wales. Some still work on runs to Leicester, Nottingham and Sheffield, among other routes, but they are outnumbered by newer trains. All the coaching stock has been extensively refurbished to individual companies' styles.

BREL Networker

Type of Train:
Diesel/Electric Multiple Unit

Class Types:
165/166/365/465/466

During the 1990s several companies – including ABB, BREL and GEC-Alsthom – built units based on the Networker design.

The diesel units are powered by Perkins engines coupled to Voith transmissions. Electric models have motors from Brush or GEC-Alsthom. Some of the third-rail Class 465 units have had their original electrical equipment replaced with units supplied by Hitachi.

The name is derived from the original route on which the class operated: 'Network SouthEast' services from London. While they are broadly similar designs, the 165 and 166 sets are slightly wider; this is because the Great Western lines they serve have a more generous loading gauge.

BREL Outer Suburban EMU

Type of Train:
Electric Multiple Unit (EMU)

Class Types:
317/318

Description

These Outer Suburban units were built at York in the early 1980s. The Class 317s had four cars and were built for the electrification of the Bedford to St Pancras line (gaining the nickname BedPan sets in the process), while the Class 318s were three-car sets for use around Glasgow.

Each of these AC units has one motor carriage. The Class 317s are powered by four GEC traction motors, while the 318s have motors supplied by Brush.

All units have since been refurbished and received modifications, the most noticeable being the removal of the outer corridor connections on the 318s.

BREL Sprinter

Type of Train:
Diesel Multiple Unit (DMU)

Class Types:
150

The Class 150s were built by British Rail Engineering Limited (BREL) at York. There were two prototypes entering service in 1984 with the production examples following a year later.

All three main variations now have Cummins engines and Voith hydraulic transmissions powering one bogie on each car. The original production batch had the same style front as the prototypes while the second batch was fitted with corridor connections.

150 001 and 150 002 were built as three-car units even though all production models only had two cars. When a three-car set has been required it has been achieved by inserting one of the later series cars in the centre, complete with redundant driving cab.

BREL Wessex Express

Type of Train:
Electric Multiple Unit (EMU)

Class Types:
442

BREL built these units in the late 1980s at Derby with components salvaged from the powerful Southern Region 1966-built 4-REP units.

Each five-car set has one motor coach with four English Electric motors with power supplied from the third rail system giving a top speed of 100 mph.

The trains were built for London Waterloo to Bournemouth and Weymouth express services but this work has now ceased. After refurbishment the sets are now used on London Victoria to Gatwick Airport and Brighton services. During the work the buffet cars were removed in favour of extra seating.

British Rail 1961 Diesel Locomotive

Type of Train:
Diesel Locomotive

Class Types:
52

Description

These locomotives were constructed by British Rail in their Swindon or Crewe works and were all given two-word names starting with 'Western' as their main operating area was the West Country.

They have two twelve-cylinder Maybach engines and Voith hydraulic transmissions and are capable of 90 mph. The engines were also used to power boats in marine installations.

Seven of these locomotives have been preserved, and those in working order can be found operating on preserved railways. One example, D1015 Western Champion, has been certified for the main line and used on enthusiast specials and charter work.

Brush 1957 Diesel Locomotive

Type of Train:
Diesel Locomotive

Class Types:
31

Brush built these locomotives as part of a pilot scheme to replace steam on the British Railway network.

An English Electric engine generates electricity for Brush DC traction motors which unusually powers only the front and rear axles on each bogie, with the centre axle being unpowered.

Once numerous on both passenger and freight work, all have now been replaced with newer locomotives. Those still on the main line can be found powering engineering and test trains for Network Rail.

Brush 1962 Diesel Locomotive

Type of Train:
Diesel Locomotive

Class Types:
47

These locomotives were built to a standard design by either Brush at Loughborough or British Rail at their Crewe Works, although many variations have been produced depending on the duties to be carried out.

A twelve-cylinder Sulzer engine powers Brush traction motors. There were 512 of these engines constructed so the locomotives were a very common sight on the railways in their heyday.

Only a few remain in front line use, and these can still be found hauling passenger trains as well as freight.

Brush 1989 Diesel Locomotive

Type of Train:
Diesel Locomotive

Class Types:
60

Description

One hundred of these locomotives were assembled by Brush at Loughborough to replace the variety of earlier locomotives still in use on freight work.

A Mirrlees Blackstone engine produces electricity via a Brush alternator which is then rectified to power six Brush DC traction motors.

Used almost exclusively on freight traffic, the class saw a much reduced workload when the new General Motors locomotives started to arrive. At one time almost the entire class was out of use, but some are now being refurbished and returned to the rails.

Brush 1993 Electric Locomotive

Type of Train:
Electric Locomotive

Class Types:
92

These electric locomotives were constructed at Loughborough by Brush to haul freight directly to Europe through the Channel Tunnel.

The Brush 1993 Electric Locomotives can run on either AC overhead or DC third rail which supplies electricity to two ASEA Brown Boveri three-phase traction motors. They have very complex electrical systems.

The 46 locomotives were originally shared between EWS, Eurostar and the French railway SNCF. Eurostar was the first to sell on their units due to the cancellation of the planned Channel Tunnel sleeper service for which the engines were intended.

24

Brush 1997 Rebuilt Locomotive

Type of Train:
Diesel Locomotive

Class Types:
57

Originally, these locomotives were part of the 512 Brush or British Rail-built locomotives shown on page 46. They were extensively re-engineered by Brush at Loughborough and returned to service from 1997.

The locomotives now have a General Motors twelve-cylinder two-stroke engine driving Brush traction motors. While the engines were new, the design of them dated back to the 1960s.

A number of these locomotives are fitted with large fold away 'Dellner' couplers for rescuing failed Pendolino and Voyager trains as normal locomotive couplers are not compatible.

25

English Electric 1952 Diesel Shunter

Type of Train:
Diesel Shunting Locomotive

Class Types:
08/09

Description

The design of English Electric's Diesel Shunters incorporates the best ideas from the British Railways predecessors.

The majority of these locomotives have English Electric engines, although some early models were fitted with Crossley or Lister-Blackstone designs. The standard versions have six-cylinder English Electric engines coupled to a pair of EE506 traction motors producing 350 hp.

With almost 2,000 similar examples built since their introduction in 1952 it is hardly surprising so many still exist, although many have been updated in some way. However, only those re-geared to provide a higher top speed have been reclassified as Class 09.

English Electric 1957 Diesel Locomotive

Type of Train:
Diesel Locomotive

Class Types:
20

Description

These locomotives were assembled at either English Electric's Vulcan Foundry or by Robert Stephenson and Hawthorns at their Darlington works.

An eight-cylinder English Electric 1,000 hp engine powers English Electric's own DC traction motors giving a top speed of 75 mph.

In a departure from the usual practice, the locomotives were made with a cab at only one end, the long narrow bonnet making them resemble a steam engine. As a result they are normally found running in pairs with the cabs at the outer ends.

English Electric 1958 Diesel Locomotive

Type of Train:

Diesel Locomotive

Class Types:

40

Most of these locomotives were constructed at English Electric's Vulcan Foundry, the remainder being assembled by Robert Stephenson and Hawthorns in Darlington.

A 16-cylinder English Electric engine provides power for English Electric DC traction motors. The turbocharger on the engines is renowned for its loud whistling noise.

None of these locomotives are now in regular revenue-earning service, however a small number have been preserved and can be seen hauling special trains on the main line.

28

English Electric 1960 Diesel Locomotive

Type of Train:
Diesel Locomotive

Class Types:
37

These locomotives were designed for mixed traffic operations. The Class 37s were purchased by British Rail as part of the modernisation plan.

Originally Class 37s had engines, generators and traction motors built by English Electric. Some were refurbished, receiving Brush or GEC alternators in place of their generators.

Most were altered considerably during refurbishment and a few members of the class were rebuilt with either Blackstone or Ruston engines. These were experimental units for the proposed Class 38, and although nothing came of the idea a larger version of the Mirrlees Blackstone MB275T engine was used in the Class 60.

29

English Electric 1962 Electro-Diesel Locomotive

Type of Train:
Electro-Diesel Locomotive

Class Types:
73

Description

The first six of these locomotives were built by British Rail at their Eastleigh works while the remainder were constructed at the Vulcan Foundry by English Electric.

Power comes either directly from the DC third-rail system or from an onboard English Electric diesel engine generating current for the English Electric traction motors.

The locomotives have spent their entire lives around the Southern third-rail network, apart from a brief spell when a few were transferred to the Liverpool area for use on engineering works. Those remaining in service are mostly to be found on freight duties, although they can still occasionally turn up connected to passenger trains.

30

English Electric Deltic

Type of Train:

Diesel Locomotive

Class Types:

55

The English Electric Class 55s were built to replace the Eastern Region express steam locomotives.

This class took their name from their engine, the Napier Deltic, a successful lightweight unit developed for marine applications. The locomotives had a pair of these 1,300 hp two-stroke engines, each with 18 opposed pistons positioned in a triangular – or delta – shape, hence the name.

Six Deltics have survived into preservation. These are numbers 55 002, 009, 015, 016, 019, 022 as well as the prototype DP1 'Deltic' which is on display at the National Railway Museum. 'Deltic' was never actually owned by British Rail and did not carry a fleet number when in use. Some Deltics can be seen in use on preserved railways and also occasionally on the main lines hauling specials or being used on rail tours.

31

GEC-Alsthom TGV-TMST Eurostar

Type of Train:
Electric Multiple Unit (EMU)

Class Types:
373

Description

The Eurostar, or TransManche Super Train (TMST), was designed for services from London to Paris and Brussels. These trains form part of the 'Train à Grande Vitesse' (TGV) family of electric multiple units that were already in service on France's high-speed network.

Each unit is divided into two 'half sets', each with their own power car. The first bogie on the first carriage is also powered. Eurostar trains are able to collect electricity from either DC or AC overhead and have provision for third-rail DC collection, which supplies power for their Brush or GEC-Alsthom traction motors.

The 'Three Capitals' units have ten carriages each, making up a 'half set', which run in pairs so that each train is 20 cars long. However, shorter eight-carriage 'half sets' were produced for services in the north of London but they never commenced; these eight-carriage trains can now be found on French internal services.

32

Description

General Electric InterCity 225

Type of Train:
Electric Train Set

Class Types:
91

The electrification of the East Coast Main Line saw the introduction of these InterCity 225s in the late 1980s.

A set included nine passenger coaches, a driving trailer and a GE electric locomotive. Each of the 31 locomotives has four General Electric traction motors which were intended to have a top service speed of 140 mph (or 225 kph) – hence the name InterCity 225.

The coaches and driving trailers were assembled at the Metro-Cammell works in Birmingham. The locomotives were constructed by BREL at Crewe Works, which served as a subcontractor for General Electric. The locomotives have a streamline cab at the front and a vertical cab at the rear, adjacent to the coaches. If a fault occurs, the locomotive can operate at reduced speed with flat-end leading, but this happens very rarely.

33

General Electric PowerHaul

Type of Train:
Diesel Locomotive

Class Types:
70

The General Electric PowerHaul freight locomotives are assembled in Erie, Pennsylvania and were first introduced by Freightliner in 2009.

The locomotives have General Electric sixteen-cylinder twin-turbo engines with GE's AC traction motors. They are highly fuel efficient and produce low exhaust emissions.

These locomotives are transported by Jumbo Ship to Newport in South Wales. Their extra power has enabled longer and heavier trains to be operated on lines that share tracks with fast passenger trains without getting in the way.

34

Description

General Motors JT42CWR

Type of Train:
Diesel Locomotive

Class Types:
66

In 1985, Foster Yeoman introduced four General Motors locomotives redesigned to fit British loading gauge and designated JT26CW-SS. This Class 59 bodywork design was utilised for the JT42CWR when introduced by English Welsh and Scottish Railways (EWS) in 1998.

The JT42CWR is fitted with an EMD 710 engine powering six GM-EMD electric traction motors. The locomotives are built in Ontario, Canada and arrive in the United Kingdom by sea in a Jumbo Ship, landing at Newport, South Wales.

A large number of similar locomotives were introduced to the British railway network between 1998 and 2008. Some of the locomotives that were originally intended for operation in Britain are now abroad, working for Freightliner and Euro Cargo Rail in places like France and Poland.

35

Hitachi Javelin

Type of Train:
Electric Multiple Unit (EMU)

Class Types:
395

These units were supplied by Hitachi from their 'A Train' family of designs. They were purchased specifically to operate the Southeastern high-speed services to Kent utilising the Eurostar tracks from St Pancras International. They commenced operation during 2009.

The six-car units consist of four motor coaches and two driving trailers. They take power from the AC overhead on the high-speed section and then switch to the third-rail DC system on the domestic lines.

With 16 traction motors per unit the trains are capable of very fast acceleration and a top speed of 125 mph – a speed that can only be reached on high-speed lines. A specially constructed depot to service the Javelins has been built at Ashford.

36

Hunslet Suburban EMU

Type of Train:
Electric Multiple Unit (EMU)

Class Types:
323

These units were constructed by Hunslet Transportation Projects and entered service around Birmingham and Manchester in 1992.

The three-car units have a centre trailer car, which houses the transformer and control equipment, and both driving cars have four Holec DMKT 52/24 traction motors.

After some problems in the early days, the sets have given good, reliable service. Many have received a mid-life refurbishment, with new flooring and seating in most cases.

37

Leyland Pacer

Type of Train:
Diesel Multiple Unit (DMU)

Class Types:
142

Description

The four-wheel underframes for these units were constructed in Derby by BREL. The bodywork was designed by British Leyland and modelled on the Leyland National bus, although the cab area was re-designed and strengthened for railway use.

When they were new, the Pacers had two Leyland TL11 engines and a self-changing gears transmission. This power plant has since been replaced by a Cummins engine and Voith gearing.

Originally fitted with standard bus-style seats, these units were the ultimate in economic railway operation. The 'wheel at each corner' design gave quite a rough nodding ride and tight curves caused the flanges to squeal badly, so much so that the type was taken out of Cornwall completely. Many have since been refurbished with new floors and better-quality seating.

38

Leyland Super Sprinter

Type of Train:
Diesel Multiple Unit (DMU)

Class Types:
153/155

These units were assembled at Workington in the late 1980s using Leyland National bus bodywork sections mounted on a steel underframe riding on twin-axle bogies.

Each carriage has a Cummins engine with Voith hydraulic transmission. The cab ends have corridor connections to enable multiple working and the units will join to other Sprinter train types.

All Leyland Super Sprinters started life as two-car units, although all except seven were converted into single-car units by Hunslet-Barclay in the early 1990s. The second cab was shoehorned into what was the centre of the train and is noticeably more cramped.

39

Met-Cam Ex-Tube Train

Type of Train:

Electric Multiple Unit (EMU)

Class Types:

483

This design of tube trains entered service on the London Underground in 1938 on the Northern Line. They are currently the oldest trains in regular daily service.

Two motors power each car of this '1938 stock', which has been converted to run on standard Southern Region three-rail DC installations, from its original four-rail London Underground configuration.

These units were converted in the late 1980s and early 1990s for use on the Isle of Wight 'Island Line'. The Met-Cam Ex-Tube trains replaced even older London Underground stock that had been responsible for introducing electrical services to the island in 1967 to replace steam-powered transportation.

40

Met-Cam Super Sprinter

Type of Train:
Diesel Multiple Unit (DMU)

Class Types:
156

Metro-Cammell's Super Sprinter was assembled in Birmingham and first introduced in 1988. These units were brought in to replace old locomotives and coaches on long-distance cross-country services as part of the Regional Railways investment programme as well as open up other routes.

These 75 mph units have two cars each with a Cummins engine, Voith transmission and Gmeinder final drive – a standard configuration for all other Sprinter types due to the reliability of the design. The trains all have two-abreast seating with most seats situated around tables in groups of four. Each two-car set also has one toilet located at an outer end.

Most Met-Cam Super Sprinters have been downgraded to operate local services as newer trains have been brought into service. Many have received refurbishments to update the original specification.

41

Parry People Mover PPM60

Type of Train:

Single Car Unit

Class Types:

139

After successful trials with a PPM50 (Class 999) two PPM60s entered service on the Stourbridge Junction to Stourbridge Town branch line in 2009.

The two units have a Ford 2.3 litre Liquefied Petroleum Gas (LPG) engine spinning a 500 kg flywheel which is then used to power the vehicle.

PPM60s are smaller than most trains by quite a margin: only seating 25 with space for a further 35 standing. However, they are ideal for the short branch they work on, which is less than a mile in length.

42

Peppercorn A1

Type of Train:
Steam Locomotive

Class Types:
LNER A1

The last of the A1 Pacifics, designed by Arthur H Peppercorn, were scrapped in the mid 1960s, having been in service since the late 1940s. The A1 Steam Locomotive Trust undertook to build another, and 'Tornado' is the result. Completed in 2008, she is now a regular sight on the main line.

Tornado is coal fired and has three cylinders. The maximum boiler pressure is 250 lb per square inch. It has six foot eight inch tall driving wheels and is coupled to an eight-wheel tender, which is a brand-new construction.

Tornado is not, strictly speaking, a replica. Improvements and compromises had to be made during construction due to advanced techniques and higher safety requirements. These include an all-welded boiler made in Germany and a steel firebox.

43

Description

Pressed Steel Suburban

Type of Train:
Diesel Multiple Unit (DMU)

Class Types:
121

These units belong to the first generation of British Railway's DMU fleet introduced from the 1950s. Vast numbers were built but by the turn of the century the majority had been withdrawn. The single-car versions were constructed in the early 1960s for Branch Line Services.

These units have mechanical transmission and British United Traction (BUT) supplied the engines. BUT was a joint venture between Leyland and AEC.

These 'slam door' units, now fitted with central door locking, can be found on Aylesbury to Princess Risborough and Cardiff Queen Street to Cardiff Bay services. Some of these units also survive in departmental use for route learning, among other duties.

44

Description

Siemens Desiro

Type of Train:
Diesel/Electric Multiple Unit

Class Types:
185/350/360/380/444/450

These versatile units are produced by Siemens and built in Germany. The first versions arrived in the UK in 2003. The 'Desiro' name was bestowed upon these trains by Siemens themselves and applies to the whole family of trains of this design.

In the diesel units, each carriage has a Cummins engine and Voith hydraulic transmission. The electric version can take power from AC overhead wires or DC third rail which then powers their Siemens traction motors.

The internal specification of the trains varies according to intended use: both local and long-distance services being undertaken. This varies from inner-London suburban work to long TransPennine services.

TRAIN TALK

Some of the terms in this book are specific to the railway industry.
The following may help to explain:

Multiple Unit: A set of trains, varying from just one to many carriages, that can be joined together to make up a longer train and driven from one cab.

Locomotive: Also called Prime Movers. A self-contained source of propulsion used to move a train. Some can be controlled remotely at the rear of a train.

Carriage: Any unpowered passenger vehicle that forms part of a train. They can be used in a multiple unit set or operated with a locomotive.

Overhead: Wires that can be found above head height which carry the electrical supply for a train to collect using a pantograph.

Pantograph: A device that can usually be found on the roof of an electric locomotive or unit to draw current from the overhead.

Third Rail: An additional rail that runs alongside the main track and provides a DC electrical supply, collected by a skate running along its surface.

ALSO IN THE 'KNOW YOUR' SERIES...

Know Your Buses

James Race
James shows 44 models. He selects those which are the most popular among bus operators as well as a few of the models less commonly seen.

Know Your Combines

Chris Lockwood
Chris shows a representative sample of the combine harvesters which are most likely to be seen working in Britain's fields.

Know Your Tractors

Chris Lockwood
Chris shows a representative sample of the popular tractor makes and models which are most likely to be seen working in Britain's arable fields.

Know Your Trucks

Patrick W Dyer
Patrick has selected 44 examples to show the wide range of truck types that you are likely to see on British roads.

 Old Pond PUBLISHING LTD

To order any of these titles please contact us at:
Phone: 01473 238200   Website: www.oldpond.com
Old Pond Publishing Ltd, Dencora Business Centre, 36 White House Road, Ipswich, IP1 5LT, U